Montserrat
Official Guide

Table of contents

To visit Montserrat

To visit Montserrat is to know one of the most important symbols of Catalonia.

To visit Montserrat is to climb a mountain and admire its beauty.

To visit Montserrat is to come into contact with the history of the Catalan people.

To visit Montserrat is to follow the itinerary of a shrine step-by-step.

To visit Montserrat is to be received by a Benedictine Monastery.

To visit Montserrat is to know one of the most important symbols of Catalonia.

To visit Montserrat is to climb a mountain and admire its beauty.

To visit Montserrat is to come into contact with the history of the Catalan people.

To visit Montserrat is to follow the itinerary of a shrine step-by-step.

To visit Montserrat is to be received by a Benedictine Monastery.

We welcome

and may the God of

But that is not all. Why not?

Because Montserrat is much more than a set of historic stones.

For many believers Montserrat is a place of encounter with God.

For many men and women Montserrat is a place of reconciliation and brotherhood.

To visit Montserrat, then, is much more that just taking a trip to the mountain.

To visit Montserrat is to come into contact with a monastery and a shrine, a place of prayer, solidarity, and celebration.

You who come to Montserrat, who may be first-time visitors, share in the beauty and life you find here, and respect them.

you,
peace be with you

The mountain

Location

Montserrat is a mountain located to the right of the Llobregat River, between the Plain of Bages (*Pla de Bages*) and the Coastal Depression. It is right in the middle of the precoastal mountain range, where it clearly stands out. The highest peak is St. Jerome (Sant Jeroni), at 1,236 meters above sea level. The Monastery is located 725 meters above sea level.

The mountain is some 10 kilometers long and 5 kilometers wide, with a perimeter of close to 25 kilometers. The surface area, however, barely reaches 45 square kilometers.

The mountain appears to be higher than it really is because it rises up abruptly from the Llobregat River to the peak. There are no other mountains nearby that are close to it in height, so that it is alone in this part of the Catalan precoastal range. The impact of this can be imagined by the fact that in just three kilometers there is a difference of 1100 meters in altitude.

Administratively, parts of the mountain belong to the towns of Collbató, Bruc, Marganell, and Monistrol de Montserrat, all of which are in the province of Barcelona. The mountain covers territory belonging to three counties, Anoia, Bages, and Baix Llobregat.

Geology

The word Montserrat means 'serrated mountain'.

The geological origins of the mountain are sedimentary. At the end of the Mesozoic Era, what today is the base of the mountain was the delta of a river that came from the Balearic continent and flowed into a large lake in the center of what today is Catalonia. At the bottom were the sediments. The submerging of the Balearic continent left the lake dry, and the delta was left uncovered and formed a large, pasty mass of conglomerates.

This mass was quite vulnerable to external agents. Movements of the Earth's crust, changes in climate and erosion worked this exposed mass some 10 million years ago and for thousands of years, to form the abrupt cliffs and rounded blocks, separated by vertical and narrow, oddly shaped channels that inspire us today. The agents also created caves and crevices in the middle of the mountain.

The rocks at Montserrat consist of a conglomerate of hardened, eroded rocks with limestone, which makes them different from the other mountains in the area, as they are unusually hard and able to withstand erosion.

The climate

Montserrat has a damp Mediterranean climate. Such pointed topography on the mountain, with abrupt changes in altitude, make for sharp contrasts in climate over small distances; it is perhaps better to talk of micro-climates as opposed to one overall climate.

According to the weather station at the monastery, located at 740 m. in altitude, average yearly rainfall is 678 l/m². The average yearly temperature is between 13°-14°C, and the influence of the sea winds is notable, as there is often thick fog, which provides generous humidity for the plants. Snowfall is rare, but every once in a while it snows considerably.

Vegetation

The plant life is a true reflection of these differences. Even if we only take into account the climatic differences, we see the presence of plants characteristic of very dry areas next to plants from very damp zones.

The vegetation on Montserrat is generally Mediterranean. Oaks (*Querqus ilex*), with the various types of underbrush, shrubs and smaller plants, cover most of the mountain. According to botanists, this is one of the best oak forests in Catalonia and even in southern Europe; due to the diversity and complexity of the plants, it is considered one of the extreme types of dry subtropical forest. There are other types of oaks (*Quercetum pubescentis*) and some pine trees (*Pinus halepensis*/*Pinus nigra*) on the mountain as well; over 1,250 species of plant grow here.

Fauna

The fauna found on the mountain is typical of the Mediterranean. The geographic diversity and botanical richness allow for abundant, varied animal life. We will only mention a few here. There are many types of birds and bats, as well as geckos. The best known mammals are squirrels, cats, and boars, and wild goats have recently been reintroduced. There are also salamanders and vipers.

Man and the Mountain

Ever since there were the first steps of an excursion, man has paid attention to Montserrat. The peak of Montgròs was climbed for the first time in 1880, and then more paths in the area were opened. Climbing with ropes has gained importance over the years: the peak of Echoes was reached in 1922, and then other major peaks such as *Cavall Bernat* (1935), or the Cylinder (1936) and St. Jerome's Wall (1948) and the Devil's Wall (1955) soon followed. Today all the major peaks have been climbed, with the concomitant detriment to the flora and fauna. There are also some major underground caves, such as *Costadreta* ("Right side"; -125 m.) and *Pouetons de les Agulles* ("Well of the Needles"; -140 m.).

Both the presence of so many people over the past few decades, and the devastation of forest fires, have had very negative effects on the conservation of the plant life and on the animal population.

The Generalitat de Catalunya, the Catalan government agency, declared the mountain a natural park.

Prehistory of Montserrat

During prehistoric times several caves on the mountain were inhabited. The two that have left us the most are the Big Cave ("*Cova Gran*") and the Cold Cave ("*Cova Freda*"), in territory belonging to the township of Collbató to the southeast of the mountain. When the explorations that had been done were first published in 1925, they became famous for being the sites of Neolithic pottery decorated with sea shells. It was the first time pottery from this period had been discovered in Catalonia, and it gave rise to the term of "Montserrat pottery". In addition to pottery, instruments made out of flint, especially knives, and pointed tools made from goat bones and animal remains (goat, sheep, dog, horse) were found. In land belonging to the township of Bruc some Neolithic burial grounds with funeral materials were also uncovered. Taken together, these findings point to an old stage of Neolithic times, going as far back as a little before 4000 BC and continuing on into 3000 BC This was when the caves were most inhabited. There are also some more recent pieces of evidence, from the Bronze Age. Some Iberian pottery and a burial site were also found in the Cold Cave. All this material is stored at the monastery, in the hopes that one day it will be put on public display.

History of Montserrat

880
The image of the Mother of God is found, according to legend, in a cave in the mountain.

888
First documented mention of Montserrat.

9th century
The four chapels in existence at the end of the 9th century, St. Mary's, St. Iscle's, St. Peter's and St. Martin's, were probably inhabited by hermits. Today only one of these, that of St. Iscle, is left, in the monastery's garden.

945
Document testifying the foundation of the monastery of St. Cecilia at Montserrat.

1025
Oliba, Abbot of Ripoll and Biship of Vic, founds the Monastery at Montserrat.

12th - 13th centuries
New Romanesque church and carving of the current image of Mother of God, which is venerated in the basilica today.

1221
Canticles of Alfonso X the Wise: initiation of the spread of the miracles of the Mother of God, and pilgrims start coming to Montserrat

1223
Institution of the Brotherhood of the Mother of God of Montserrat

1223
First accounts of the presence of a Boy's Choir, the first in Europe.

14th century
Montserrat starts to spread throughout Europe.

1409
The monastery becomes an independent abbey.

1490
A printing press is installed at Montserrat

1476
The Gothic cloister is built.

1493
Bernal Boïl, a hermit from Montserrat, goes with Christopher Columbus on voyage to America. One of the islands in the Antilles is named Montserrat. Initiation of the spread of worship of the Mother of God of Montserrat in the Americas.

1522
Ignatius of Loyola makes a pilgrimage to Montserrat.

1592
The present-day church is consecrated.

16th - 18th centuries
Many cultural activities are started at the monastery.

17th - 18th centuries
Importance of the music school at Montserrat.

17th - 18th centuries
Important foundings of dependencies in Bohemia and Austria.

1811-1812
Montserrat is destroyed by Napoleon's army.

1835
Land acts, due to which the monastery loses all of its property and only one monk remains

1844
Monks return to Montserrat.

1858
Under the leadership of Abbot Muntades, the definitive reconstruction of Montserrat is undertaken.

1880
Millenary celebrations at Montserrat.

1881
Celebration of the Coronation of the Image and Proclamation of the Virgin of Montserrat as Patron Saint of Catalonia

1901
Inauguration of the new façade of the basilica.

1915
I Liturgical Congress at Montserrat.

1931
Celebration of the 11th Centenary of the founding of the monastery.

1936-1939
Spanish Civil War. The monks have to leave the monastery; during the war 23 monks are killed. The autonomous government of Catalonia saves Montserrat, and frees it from looting and destruction.

1939
Once the war is over, the monks return.

1942
The first stone of the new façade of the monastery is placed.

1947
Celebration of the Enthronement of the Image of the Mother of God, marking the first movement of civic reconciliation in the country after the Civil war.

1959
The main altar of the basilica, facing the liturgical assembly, is consecrated and the rectory and the monks' choir are remodeled.

1965
II Liturgical Congress at Montserrat.

1968
The new façade is completed.

1970
The monastery, with 300 intellectuals locked inside calling for respect for human rights by Franco's dictatorship, is put under siege by the police for 2 days.

1976
The new complex, with restaurants for the large numbers of pilgrims at the entrance to the shrine, is inaugurated.

1980-1981
Celebration of the 100th anniversary of the Coronation of the Image and of the Proclamation of Montserrat as the Patron Saint of Catalonia

1982
The new museum of modern Catalan painting is inaugurated.

1982
Pope John Paul II, with the Church of Catalonia, visits Montserrat.

1986
On August 18 a major forest fire devastates a large part of the mountain.

1987
The Generalitat de Catalunya makes the mountain a natural park.

1988
The project to restore the basilica is undertaken.

1990
III Liturgical Congress at Montserrat.

1991
Restoration of the basilica is begun on September 16.

1992
Celebration of the IV Centenary of the consecration of the basilica.

1995-1996
Inauguration of the exterior and interior restoration projects for the basilica.

1997
Inauguration of the restoration of the Holy Grotto on March 19.

1997-1998
Celebration of the 50th anniversary of the Enthronement of the Image of the Mother of God of Montserrat.

Montserrat Today

What is Montserrat?

Montserrat is, first of all, pilgrims, visitors, and tourists that come here to worship the image of St. Mary in the basilica.

Each year thousands of people come to Montserrat. To attend to their most immediate needs, next to the shrine lodging and restaurants have been built. Some 250 people work here, following the guidelines set by the monastery.

From its beginnings, in addition to the mountain and the shrine with its holy image, two other institutions have marked the history of Montserrat: the boys' choir and the Monastery.

The Boys' Choir

Today there are about 50 boys in the choir. In addition to receiving a strong intellectual education, they are taught music. In the past few years the Choir has also given concerts outside Montserrat and has traveled to foreign countries. It has recorded over 100 albums and compact discs. But the main purpose of the Choir still is, as it has always been, to participate in the liturgical celebrations and community prayer services held in the basilica.

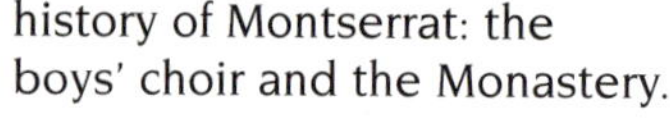

Every day at 1:00 p.m. the boys sing the *Salve and Virolai*, in the afternoon, after vespers, at 7:10 p.m. they sing the *Salve montserratina*, in which counterpoint and Gregorian chant are alternated with the monks' choir and a polyphonic motet. On Sundays and holy days they take part in the concelebrated mass and in vespers. The Boys' Choir is not present at Montserrat in July, Christmas holidays, and on a few other days during the year.

The Monastery, or the community of monks

The community of monks at Montserrat today, together with those at El Miracle and St. Michael of Cuixà, which administratively depend on it, consists of about one hundred Benedictine monks who follow the Order of St. Benedict (6th century). The life of the monks is devoted to prayer, work, and everything every person is. This willingness to participate is also characteristic of all those who come to the monastery. This is where the monastery gets is unique character of shrine from, a character centered on the monks' prayer, open to everyone, to welcome those who come to Montserrat.

This way of life means that the monks are men of inner silence willing to listen to the voice of God as manifested in the Holy Scriptures and in the developments of the Church and the world. This sense of the presence of God in one's own life, a fundamental trait of monastic spirituality, is what one ultimately perceives as different about Montserrat,

and is what makes sense out of all the activities of the Church and of society.

The monks work at many things. Some are involved with the internal organization of the monastery, others are researchers or teachers in a variety of fields (history, theology, Bible studies, liturgy, music, etc.). We should note the importance of the publishing house run by the monastery.

Others serve the shrine: they attend to the pastoral needs of the basilica, receive the various groups, direct retreats, make speeches, take care of the guests, or in general deal with the guidelines the monastery wants to establish for all of the services given to pilgrims and visitors to the shrine and the park.

Finally, some monks, along with lay teachers, are involved with teaching the boys in the choir. Some monks are particularly interested in musical direction, composition, or research, and have become well-known musicians (especially organists). Given the importance of music at Montserrat, the monastery also produces recordings.

A single community

Every day, and at several times during the day, the eight bells, cast between 1955 and 1958 and located on the Gothic bell tower (14th century), ring depending on the celebration. The largest bell, weighing seven and a half tons, was placed there in 1958 and is known as St. Mary's Bell. The bells bring the monks, the boys in the choir, those who work at the shrine, pilgrims, visitors, and tourists all to the basilica to form a single community in prayer, which, despite the many languages, looks to God the Father, source of all goodness, to ask for solidarity among people and peace on Earth.

The main area of the shrine

ITINERARY I

The itineraries we present are conceived so that you can visit Montserrat without a guide. The time indicated is that necessary to go to the place proposed. Please remember that the most important part of visiting Montserrat is not enjoying the architecture, but rather discovering the religious, cultural, social, historical, and ecological values that together symbolically express the life of a people.

We begin this tour at the Square of the Cross (*Plaça de la Creu*; 706 m.), which is named after the cross (1962) to the left of the square, a monument to St. Michael, the patron saint of the mountain of Montserrat. As we get closer, we see the meaning of the saint's name, 'Who is like God?,' engraved on the cross. The monument is the work of the Catalan sculptor Josep M. Subirachs (b. 1927). To the right of this square is the Information Office, where you may ask questions and obtain the free brochures that, along with this guide, will make your stay at Montserrat informative and enjoyable.

Abbot Oliba Square

Going down from the Plaça de la Creu and continuing along the *Pujada dels Roures* (Oak Hill) we arrive at the *Plaça de l'Abat Oliba* (Abbot Oliba Square). Weather permitting, we will have been able to rest peacefully under one of these trees, which are widespread in this area. Once in the square, we see that it is surrounded by three large buildings where pilgrims stay in rooms or apartments equipped with modern facilities. The ground floors of these buildings, especially that of the oldest one of the three, known as *Nostra Senyora* (Our Lady), built 1895-1904 and designed by the architect

see a majestic bronze sculpture (1992) dedicated to Abbot Oliba, who founded the monastery in the year 1025. The sculpture is by Manuel Cusachs (b. 1933). The Abbot, who was the Bishop of Vic (county of Osona), sits on a chair on the back of which the

Francesc de Paula del Villar i Carmona (b. 1860- d. 1927), contain basic services such as grocery store, souvenirs, bookstore, bar, doctor's office, bank, police, etc. At the end of the square there is a 4-spouted fountain where we can drink and refresh ourselves. Right above that, in a landscaped area continuing from the Torrent de Santa Maria (St. Mary's Stream) that flows down the mountain, we

bell towers of the monasteries of Ripoll (in Osona) and Sant Miquel de Cuixà (the Conflent area, in France), of which he was also Abbot, are represented. His left hand holds the plans for the first church at Montserrat, and his right hand welcomes everyone, who comes to the monastery and shrine at Montserrat, following the Benedictine tradition.

In the middle of the square we see hundred-year old cedar trees (*Cedrus libani*) which were brought here from the mountains in Lebanon by P. Bonaventura Ubach (there is more information about this monk in Itinerary II: the Museum), and a beautiful (*Taxus baccata*), a tree native to central Europe found in the shady, damp areas of the mountainside.

The area surrounding these trees has been landscaped and there are stone benches for you to sit on. The square reminds us of the squares of our towns, and at the same time provides modern conveniences while making us think of the centuries-long history lived by men and women who, like us, have rested and refreshed themselves at the fountain before entering the religious area proper.

St. Mary's Square (La Plaça de Santa Maria)

The entrance to this new area, located very near to the basilica, is from the square we are now in. To the right we can see a portion of the old wall, and in the middle we see the entrance gate with the coat-of-arms of Montserrat dating from 1565. Through this Gothic gate we walk up the *Pujada de Nostra Senyora* (Our Lady's Hill), decorated by a row of colourful magnolias. Halfway up, in a niche on the wall to the left, there is a sculpture of St. George (1986) by Josep M. Subirachs. Many visitors pause because of the way it looks at you, as it seems to follow you no matter where you go, and because it is reminiscent of some of the sculptures on the Façade of the Passion of the Temple of the Holy Family (*Sagrada Família*) in Barcelona, which is the work of the same sculptor.

As we go up we see the large esplanade carved out of the mountain that forms the *Plaça de Santa Maria* (720 m.). It is made up of three stepped squares (1929) designed by the architect Josep Puig i Cadafalch (b. 1867 - d. 1956). When we reach the largest of the three we can appreciate the architecture of Montserrat: behind us, to our left, we see the lodging and service area for pilgrims and visitors. Right next to us, in the center of a garden, is a beautiful Gothic cross dating from the fifteenth century. And in front of us is the new façade to the monastery and the partition that closes off the square (1942-1968), which was designed by the architect Francesc Folguera (b. 1891 - d. 1960). Most of the façade is built with polished stone from the mountain, and if we look closely we can see the type of formations we described earlier in our discussion of the geological origins of the mountain.

At the top of the façade there is a phrase in Latin, 'Urbs Jerusalem Beata Dicta Pacis Visio', which means "Happy city of Jerusalem, called the vision of peace." We find this several times in the Bible, and it refers to the heavenly Jerusalem, the reference point for all Christian sanctuaries. The three large upper balconies are decorated with works of the sculptor Joan Rebull (b. 1899 - d. 1981), the one on the left evokes the figure of St. Benedictine, father of monks and patron saint of Europe; the one in the middle represents the proclamation of the Assumption of Mary as dogma by Pope Pius XII; and, the one on the right portrays St. George, the patron saint of

Catalonia, with a representation of the monks who tragically died during the Spanish Civil War (1936-39). Often, on Sundays and holidays, this square is filled with popular culture - *sardanes* (a traditional Catalan dance), *castellers* (human towers), *gegants* (giant figures), and other dances - in which many pilgrims take part after having attended the religious ceremonies held inside the basilica.

Connected to the left side of the façade we can see the

remains of the old Gothic cloister (1476) that was built by the commendatory Abbot Giuliano della Rovera, who would later become Pope Julius II. Only two wings remain, with capitals representing nonreligious themes and with the coat-of-arms of Montserrat and of the Abbot who paid for the construction. The ground floor of the building next to the Gothic Cloister houses the Shrine Office, where we can ask for spiritual guidance and other information.

The series of sculpted figures that encloses the square to the right is devoted to the founding saints of the various religious institutions that have been related to Montserrat over the centuries. This group of arches, with sculptures by Claudi Rius (b. 1892 - ?), Francesc Juventeny (b. 1906 - ?), Enric Monjo (b. 1896 - d. 1976), and Joaquim Ros i Borafull (b. 1906 - d. 1991) offered from 1949 to 1953, is an excellent balcony from which on clear days we can see the Llobregat river basin, which flows into the Mediterranean after having passed through many industrial towns near Barcelona. In the distance we can also see the Tibidabo, and the telecommunications tower designed by Norman Foster, reminders of the 1992 Olympic Games.

The atrium of the Basilica

If we cross one of the five arches leading to the atrium of the shrine, we can see to our left a sculpture of St. Benedict (1962), a work in forged iron by Domènec Fita (1927). This sculpture shows us the entrance to the monastery, where the monks live. It is framed by a stone frieze (1960) alluding to the founding of the monastery in 1025 by Abbot Oliba, and to the legend that places the finding of the image of the Mother of God in 880. The frieze is the work of the sculptor Enric Monjo. The monastery is not open to visitors, in order to respect the monks' way of life. There is a valuable library with over 300,000 volumes that is open to scholars.

Here we can see a Gothic gate (15th century) and two tombs (16th century), which together with the Romanesque gate of the old church, in the hallway to the right leading to the atrium, and the Gothic cloister, remind us of the medieval and Renaissance Montserrat that was destroyed by Napoleon's troops (1811-1812). On the sides of the outlying halls we see representations of the visits to Montserrat made by Ferdinand and Isabelle and John of Austria. These murals (1957), which are rather naïf in style, are the work of the painter and sculptor Francesc Fornells-Pla (b. 1921). If we go through the central area, we will see two sculptures by Josep Clarà (b. 1878- d. 1958), one of St. Joseph and the other of St. John the Baptist.

The basilica's atrium, known as the atrium of Abbot Argerich (18th century) was beautifully decorated in 1952-1956 with designs by Josep Obiols (b. 1894- d. 1967) and Father Benet Martínez (b. 1918- d. 1988). Since Montserrat is a shrine and its church is a minor basilica, as named by Pope Leo XIII in 1881, the designs to the right bring to mind the most important basilicas and shrines in Christianity; those to the left offer us a brief summary of the history of Montserrat. As we visit the atrium, going from right to left, we first see the baptistry, designed in a project by the Committee for the Support of Decorative Arts (*Junta del Foment de les Arts Decoratives*) and opened in 1958, with a beautiful doorway by the sculptor Carles Collet (b. 1902- d. 1983). The upper part shows the mystery of the Church that gives the sacraments (to the right) that bless human activity and life (to the left). The inside is decorated with mosaics by Santiago Padrós (b. 1918- d. 1971) and overlooking everything is a painting of Jesus' baptism by Josep Vila-Arrufat (b. 1894- d. 1989).

Next to the baptistry is a sculpture of St. Ignatius of Loyola by the sculptor Rafel Solanic (b. 1895- d. 1990). Next to it is a stone from 1603 that reminds us in Latin that St. Ignatius spent a night vigil before the holy image of the Mother of God on March 25, 1522. If we look at the floor, we will see a small piece of black marble with an inscription in the middle of the arch; this is where the altar of the old Romanesque church was, with the image of St. Mary that today we venerate in the basilica. There are also other important sculptures around the atrium: St. Anthony M. Claret (?), also a work by Rafel Solanic; King John I of Aragon

(1956) and St. Gregory the Great (1957), by Frederic Marés (1883-1991), King Charles I, done in Marés' studio; and lastly, St. Pius X, a work by F. Bassas (?). In front of the sculpture of John I there is the eternal flame of the Catalan Language (1968).

The beauty of the black and white marble floor must also be singled out (1952). It gets its inspiration from the floor of the Capitolium in Rome designed by Michelangelo. In the center there are allegories and a Latin inscription that refer to baptism. From here the basilica appears to us in all its majesty. It was Abbot Bartomeu Garriga who in 1560 placed the first stone, but it was not consecrated until 1592, and there were still parts unfinished. On the occasion of the 400th anniversary of the consecration (1992), restoration of both the inside and outside was begun (1991-1995) under the direction of the architect Arcadi Pla i Masmiquel (b. 1945). There had been three previous restorations; for example, between 1900-1901 the old Baroque façade was replaced by the one we see today, the work of Francesc de Paula del Villar i Carmona and carved by the brothers Venanci and Agapit Vallmitjana (b. 1828 - d. 1919 and b. 1830 - d. 1905, respectively).

Once we reach here, we can do one of two things: either visit the basilica, or go directly to venerate the Image of the Mother of God by entering the side door to the right, under the arches. We suggest you visit the basilica first, so that you will have a better overall idea of Montserrat. Then we can go worship the Image of St. Mary.

We would please ask you to be silent while in the basilica, and to not walk around inside during religious acts.

The Basilica

The church at Montserrat was constructed in Gothic tradition and also includes Renaissance shapes that Catalan architecture began to incorporate at the time (16th century). As we go in, we see the front of the nave, with the monks' choir and the main altar. In the upper part of the center area there is a gilded temple, designed by Francesc de Paula del Villar, where we can see the silver throne with the Image of the Mother of God. As opposed to many other churches, this church has a single nave with chapels connecting to one another between the buttresses, and it has side platforms. The arched dome and the moldings reach downwards to form little half-circles, and where we would expect all the lines to meet we see the great octagonal dome. The basilica was severely damaged during the Peninsular War against Napoleon's France (1808-1814), and was reconstructed at the end of the 19th century. The Romanesque-Byzantine decoration of the inside has been the subject of some controversy in terms of its quality, but we can say that the architects, painters, and sculptors that decorated it were considered masters of Catalan modernism and the symbolism of the time.

As we enter we realize the church is not enormous, although if we stop to consider when and where it was built, it is quite large. The central nave is 58 meters long, 15 meters wide, and 23 meters high to the top of the dome. Today, thanks to restoration (1991-1996), the church has a lot of natural light and still maintains a sense of privacy.

There are many candles around the central nave and side chapels, some of which are important works of art and quite representative of Catalan jewelry making after the Spanish Civil War (1940). These votive lamps, in commemoration of those before Montserrat was destroyed (1811-1812), have been offered by counties, towns, and associations from around Catalonia and by Catalan associations abroad. All together they symbolize the continued presence of the Catalan people at the feet of St. Mary of Montserrat, their patron saint.

There are sculptures of the prophets Ezekiel, Jeremiah, Isaiah, and Daniel on the central pillars of the nave that remind us of the prophecies of the Virgin Mary. They were carved from wood by Josep Llimona (b. 1864 - d. 1934) and were put in place in 1896. Now we can visit the chapels to our left.

The Chapels in the Basilica

The first decorated chapel located close to the presbytery is dedicated to St. Scholastica, St. Benedict's sister. The sculptures were done by Enric Clarasó (b. 1857 - d. 1941) and Agapit Vallmitjana, and the

chapel was restored in 1886.

The Chapel of the Most Holy One was remodeled according to a project by Josep M. Subirachs in 1977. It is sober, to invite silence; it reminds us of the reforms in religious art after Vatican II (1962-1965). If we look closely we can see a large stained glass window that separates it from the nave. The railing is bronze, and there are engravings of symbols we should take note of: those on the left represent life, whereas those on the right, dealth; between the two there is a poem by the Catalan poet Salvador Espriu (b. 1913 - d. 1985) to St. George, patron saint of Catalonia, in which he asks that our country never again live a civil war but rather only peace. Once inside, and as the artist wished, we only pay attention to the important features--the altar and retable. It is built from a single piece of cement to underscore the unity it symbolizes: every time we celebrate the Eucharist at the alter we announce the victory of live over death, which is represented by a bone at the foot of the altar, because of the resurrection of Jesus, who is symbolized on the retable that rises from the altar. The design of the liturgical items also reflects this sobriety: the engraved silver tabernacle (1967) was done by Manuel Capdevila (b. 1910), and the remaining

items (1996-97), also made of silver and engraved with enameled wood, were done by Joaquim Capdevila (b. 1944).

The third chapel contains a monumental painting (1904) by Josep Cusachs (b. 1851 - d. 1908), representing the exodus to Egypt. Note the detail on each person and the choice of colors. On the one hand we see a realistic vision of the anxiety felt by the three fugitives who, with the aid of a strong mule, are running across the rocky land to exile; on the other, we see serene angels surrounded by fog and natural colors that seem to take Jesus, Mary, and Joseph to a safe place.

The next chapel is that of the Holy Christ. The image of Christ on the cross is by Josep Llimona, and was offered by the Carriers of the Holy Christ of the diocese of Barcelona in 1933.

The last chapel is dedicated to the Immaculate Conception, and was opened in 1910. The project (1906) was done by the architect Josep M. Pericas i Morros (b. 1881 - d. 1965). To understand this work it is necessary to understand the author. His point of reference was Antoni Gaudí, and later his work showed all the characteristics of Catalan

modernist architecture. This chapel is thus an excellent example of all the features of Catalan architecture from that time.

The chapels to the right are described below.

The presbytery and monks' choir

Now we go towards the presbytery. The modernist paintings decorating the walls are interesting. The large ones above are by Alexandre de Riquer (b. 1856 - d. 1920), Miquel Utrillo (b. 1862 - d. 1934), Joaquim Vancells (b. 1856 - d. 1942) and Joan Llimona (b. 1860 - d. 1926).

The two closest to the throne area have angels, and those in the middle are allegories about Montserrat and Catalonia (right) and the Church (Left); the two at the ends represent scenes from the finding of the Image of the Mother of God in the Holy Grotto. The four paintings located on the lower part are scenes related to the life of St. Mary, the work of Joan Llimona, Dionís Baixeras (b. 1862 - d. 1943) and Lluís Braner (b. 1836 - d. 1929).

The presbytery and monks' choir were remodeled in 1957-1958. The initial project was by the architect Frances Folguera, and he was aided by the monks Father Pere Busquets (b. 1925) and Father Crisòleg Picas (b. 1922). The main altar is made from a single stone weighing eight tons that was removed from the mountain. It rests on a stone from the earlier altar in the basilica, from which the material was taken for the stone sculpted by Joan Rebull with the Eucharistic symbol of a pelican. Behind this there are relics of several saints. The front enamel piece, the work of Montserrat Mainar (b. 1928), is placed in a silver frame made by the silversmith Manuel Capdevila. The altar has a gold cross with ivory, which has been attributed to Ghiberti (16th century), which was offered in 1960 by the College of Physicians of Catalonia and the Balearic Islands. The cross hands from the crown, also made by Manuel Capdevila and decorated with designs by Joaquim (Ferran) González i Cañete (b. 1932 - d. 199?), a student of Josep Grau i Garriga (b. 1929).

The entrance to the chapel of the Image of the Mother of God

To worship the image of St. Mary, which we see in the middle, we need to leave the basilica and then enter again by the side door to the right, which is clearly indicated.

The drum over the door corresponds to the former Baroque door of the shrine. Below there is an inscription that tells us of the celebrations of the 100th anniversary of the crowning of the Image of the Mother of God and the proclamation of her being the patron saint of Catalonia (1980-1981). As we go ahead we can see the side chapels on the right of the basilica. They not only contain many sculptures and stained glass windows with references to the life of Mary, but also the tombs of members of families that made the decoration of Montserrat possible at the end of the 19th century.

The first chapel is in honor of St. Peter, with a bronze sculpture (1945) by Josep Viladomat (b. 1899 - d. 1989). There is a plaque commemorating the pilgrimage of Pope John Paul II to Montserrat (1982).

The second chapel is in honor of St. Ignatius of Loyola, the most representative pilgrim to Montserrat. He is represented in the painting in the middle, by Ramir Lorenzale i Rogent (b. 1859 - d. 1917), and to our right there is a reproduction of the sword St. Ignatius left at Montserrat during his vigil on March 25, 1522.

The third chapel is the Chapel of St. Martin (1898), which

shows a decisive moment in the saint's life as his sword cuts away part of his cape to give it to a needy man; the face of the poor man is a self-portrait of the artist, Josep Llimona.

The fourth chapel has a pretty retable in the modernist style (1891), the work of Francesc Berenguer i Mestre (b. 1866 - d. 1914), in honor of St. Joseph Calasantius, who also visited Montserrat in 1785. It was offered by the *Escola Pia* in 1891.

The last chapel surprises us with an oil painting (1980) by Montserrat Gudiol (b. 1933) that evokes St. Benedict, the founder of the order, in his youth; in his hands he holds the rules of the order that has had so much influence on the history of Christianity.

The throne of the Image of the Mother of God

The idea of reforming the throne room of the Image of the Mother of God and of building a new stairway to the chapel dates from 1944, with the goal of completion being the Enthronement on April 27, 1947. Work was not completed until 1954 during the Marian Year. Several architects, artists, sculptors and jewelers took part in the project under the direction of the architect Francesc Folguera and the painter Josep Obiols, in order to create a synthesis of Catalan art from the mid-20th century.

We go through a large alabaster doorway with Biblical references that Christian tradition has related to Mary, Mother of God. It was opened in 1954. Sculpted by Enric Monjo, it is flanked by two candlesticks, also of alabaster, made by Rafel Solanic. The interior walls of the stairs are decorated with mosaics by Santiago Padrós, following a design by Father Benet Martínez.

Those on the right show saints who were virgins, and those on the left, those who were mothers.

We are now in the room where years ago the ex-vots and offers by pilgrims to the Mother of God were displayed. Today, there are only three in representation of all of them: the three flags next to the large modernist armoire made of ebony and ivory (1928) by Esuebi Busquests (b. 1875 - d. 1962).

Their presence next to St. Mary is thought to help the Catalan people live in reconciliation and brotherhood with all men and women, and in solidarity with other peoples and cultures while still maintaining its own identity. On the other side of the room there is the Fountain of the Mother of God, with relief by Carles Collet, who also did the capitals. The base of the fountain shows the miracles of Jesus that are

related to water. Water has always been a symbol of life. Mary gives us Jesus, our life. We no longer need to fear anything because he is with us, and takes care of us. He gives us this water — the baptism of the Holy Spirit — and we no longer will thirst, and we will obtain eternal life. The sculptures on the capitals show us the medieval miracles attributed to St. Mary of Montserrat.

At the end of the room a painting by Carlo Maratta (b. 1625 - d. 1713) represents the birth of Jesus, in the arms of Mary, his mother. The murals on the walls leading up to the throne of the Image of the Mother of God are by Josep Obiols, and represent virtuous, brave women mentioned in the Bible.

Doors made of repoussé silver lead to the throne, and are the work of Josep Obiols, Rafel

two alabaster candlestickes on either side of the throne, by Pere Jou (b. 1891 - d. 1964), have sculptures that bring to mind the processions of pilgrims that have worshipped at this holy image.

The throne of the Image of the Mother of God, as a repoussé silver retable, was given by the Catalan people in 1947. It has two silver reliefs, by the jeweler Ramon Sunyer (b. 1889 - d. 1963), and it is held

Solanic and Manuel Capdevila. The throne, which can be seen from the center of the church, is completely covered with Venetian mosaics designed by Josep Obiols and made by Santiago Padrós. we see Mary symbolized as the Mother of humanity, of the Apostles, of the Church, of Catalonia, of the monks and pilgrims. The nine silver lamps flanking the throne, made by Xavier Corberó (b. 1901 - d. 1981), represent the eight dioceses of Catalonia and Montserrat. The

together by molds made by the jeweler Alfons Serrahima (b. 1906 - d. 1988), following a project by Joaquim Ros i Bofarull, and represent the Nativity and Visitation of Mary. Above the image, there are angels, by the sculptor Martí Llauradó (b. 1903 - d. 1957), holding the reproductions of the crown, scepter and lily, a gift of the Catalan people to the Mother of God in 1881 (the originals, by Francesc de Paula del Villar i Lozano, Joan Suñol and Joaquim Cabot, are on

display in the Museum). The Image is a beautiful polychromic wooden carving from the 12th/13th centuries, rests on a polished stone taken from the Mountain. At the base is a sculpture of St. Michael, by Josep Granyer (b. 1899 - d. 1983). The image is protected by an oval glass. The dark color of her face and hands, which places her with the black virgins, has given rise to her popular name, la *Moreneta*. The dark color is not from the wood, which is not black, nor from earlier coverings of paint. We know from historical descriptions that it has darkened over the years. The ritual for venerating the Blessed Virgin is to kiss or touch her right hand, which holds a sphere symbolizing the universe, and open you other hand out to Jesus, her Son, who gives us eternal life.

The chapel of the Image of the Mother of God

We go down from the throne, and to the right there is a chapel located at the center of three large neo-Romanesque apses connected to the east façade of the temple. This was begun in 1876 and not completed until 1885. The work was first done under the direction of Francesc de Paula del Villar i Lozano (b. 1828 - d. 1901), aided by the young architect Antoni Gaudí (b. 1852 - d. 1926), and then by his son, Francesc de Paula del Villar i Carmona, who also worked on the restoration of the church.

The decoration inside of the chapel of the Mother of God was not finished until 1887. We can see the interesting sense of harmony of the various neo-Romanesque, neo-Gothic, and Renaissance elements that place it within the Catalan pre-Modernist movement. On the other side of the main stained glass window, right in front of the Image of the Mother of God, there is a sculpture of St. George, the patron saint of Catalonia. This work by Agapit

Vallmitjana was carved out of wood and then gilded; the facial features lead us to believe it was a portrait of his brother, Venanci Vallmitjana.

The painting on the dome (1896-1898) is stunning; it is the work of Joan Llimona. It was very difficult to do because of the problems with perspective. The light colors emphasize the glorification, with Mary, Mother of God holding Jesus in her arms, surrounded by angels and archangels; in the background there is the mountain of Montserrat seen to mark the intersection of the heavens with the earth. On the lower part of the dome there is a line of pilgrims--from the right the church figures come, and laymen come from the left. The people represented are all related to the symbolism and history of Catalonia. This dome is truly one of the artistic gems of Montserrat and of all modern Catalan painting.

The Ave Maria Path

As we leave the access room we are now outside, by the religious area continues along the Ave Maria Path (*Camí de l'Ave Maria*). This area is run by the thick temple walls and the mountain itself. To unify the inside with the outside, and at the same time protect people from the rain, the architect Josep M. Martorell (b. 1925) designed a covering (1982) out of transparent metacrylate supported by forged iron and cement that lends a sense of motion. We can still see the sky, plants, and rocks through it, as well as the 16th century façade of the basilica. Both the covering and the mountain are integrated into this place of worship of St. Mary.

Few words need to be said. It is enough to see the thousands of candles that are lit daily, the tall candles, made from the wax that did not completely burn on the other candles, the votive candles, an extension of those that light up the inside of the temple, the well-known prayers to the Mother of God written on majolica plaques, the Marian litanies of the rosary forged in iron, the bronze sculpture (1978) by Apel·les Fenosa (b. 1899 - d. 1988) evoking the angel's Annuciation to Mary, and, at the end, another image of St. Mary, as seen by Baroque artists, here reproduced on a large majolica (1982), the work of Joan Guivernau (b. 1909). We must remember one thing about this path: every flame, every plaque, every candle symbolizes a prayer to St. Mary made by a person, a family, an entire people.

The Ave Maria Path takes us back to the basilica's atrium, where we started our long, intense journey. Now we are better equipped to reflect on the meaning of our visit. The most important part was not the works of art, the artists, the dates or periods when things were done, but rather the perception of religious, human life that thousands of people experience here throughout the year. A holy place because in the central nave every day and at several times during the day the monks and boys in the

choir, pilgrims, and visitors meet together, forming a single community to celebrate and sing together with Mary that God is holy and saves us. A holy place because around this community there is a large ring of pilgrims praying for Mary to

intercede for us, so He will show us mercy and love forever. The liturgical assembly in the church fills the ring with people's devotion. The real temple of Montserrat is formed not by the rocks, but rather by the men and women who visit Montserrat and by those by live there. The words of St. Paul come to reality: "...you were God's temple and the Spirit of God was living among you? The temple of God is sacred, and you are that temple." (1C 3,16).

The Museum

Itinerary II

This itinerary complements the first. After visiting the basilica, and if you have an hour of time, we recommend visiting the museum. It is one of the most interesting museums to visit in Catalonia due to the variety and importance of the works it contains.

Most of the works of art are donations by private citizens to the monastery so that they can be displayed to those visiting Montserrat. The Monastery is happy to provide this service, with a didactic spirit of serving culture. This visit gives us the chance to know a part of the history of Montserrat and of Catalonia through art.

This pedagogical view of the museum was held by Father Bonaventura Ubach (b. 1879 - d. 1960), a monk at Montserrat, who collected a large amount of objects and archaeological, ethnological, botanical and zoological materials related to the Biblical world during his long stays in the Middle East from 1906 to 1951, in order to illustrate the Holy Scriptures and increase knowledge of them in Catalonia. As a result of his enthusiasm, the group of

monks in the Scriptorium Biblicum at Montserrat began translation of the Bible into Catalan directly from the Middle Eastern languages in 1926. They also started the museum of the Biblical East, which we suggest you visit, along with the other collections.

The museum is located under St. Mary's Square. The entrance is between the steps that go from the main square to the second square. We can see the area designed by the well-known architect Josep Puig i Cadafalch (b. 1867 - d. 1956), which is supported by an iron structure. The two main areas of the museum were remodeled in 1980-82 under the direction of Father Pere Busquets (b. 1925) and Father Crisòleg Picas (b. 1922). Some years later they were enlarged by Josep Gargante (b. 1943).

The collections on display to the public are:

• Archaeological collection from the Biblical Middle East, consisting of representative objects from Mesopotamia, Egypt, Cyprus, and the Holy Land.

• Collection of paintings from the 13th - 18th centuries, most of which were acquired in Italy between 1914 and 1920. The most representative painters are Caravaggio, Luca Giordano, A. Vaccaro, Tiepolo, Berruguete, Morales, and El Greco.

• Group of liturgical object from the 15th - 20th centuries connected to Montserrat. It shows how styles have evolved, and the various artistic techniques used by silversmiths.

• The Nigra sum exhibit is devoted to the symbols related to the Mother of God of Montserrat over the centuries. The paintings, sculptures, drawings, engravings, medallions, etc. show us the artistic taste of each period.

• A collection of Catalan painting and sculpture from the mid-19th century to the mid-20th century that is very representative, including important figures such as Rusiñol, Casas, Nonell, Mir, Picasso and Dalí. This collection is complemented by a group of French impressionist paintings, including works by Monet, Sisley, Degas, Pissarro, Rouault and Poliakoff. The presence of works by other renown artists, such as Singer Sargent, Sorolla, Zuluoaga o Julio Romero de Torres, and series of lithographs by Picasso, Dalí, Miró, Tàpies, Clavé, Le Corbusier, and Braque gives us an idea of this significant collection's commitment to contemporary painting.

Itinerary III

The Way of the Cross (Via Crucis)

(20 minutes)

Brother Garí's Viewpoint (La Miranda de Fra Garí)

(20 minutes)

St. Michael's Chapel (La Capella de Sant Miquel)

(45 minutes)

St. John via St. Michael

(1 hour 15 minutes)

This itinerary has several options: take only one of the walks suggested or complete the Way of the Cross along with another one of the routes. We do not recommend doing the itineraries one after the other because what we are suggesting here is a leisurely walk, not an excursion. The time given also corresponds to a rather slow pace. All of these points will give you scenic views of the monastery and shrine.

The Way of the Cross

The Way of the Cross is an act of devotion consisting of following, praying and meditating at 14 crosses or stations on what Jesus did carrying the cross from the house of Pilate to Calvary, concluding with the crucifixion and the burial of Jesus. For Catholics, following the Stations of the Cross is a manifestation of their will to identify themselves with the same feelings as those of Jesus Christ, and wait with him for the resurrection promised by God the Father for all those who believe in his Son.

The path that takes us to the Way of the Cross begins at Abbot Oliba Square (710 m. in altitude). To the left of the fountain there are some steps, at the end of which we see the first station. Although the route is close the shrine, it is silent and often solitary, and always shaded by oak trees. It is one of the nicest and easiest walks we can take without going far from the main area. It is a place of prayer for many pilgrims. In front of the twelfth station there is a panoramic view of the monastery and shrine we highly recommend.

Between 1904 and 1919, Enric Sagnier (b. 1858 - d. 1931), one of the most famous architects in the modernist movement, and Eduard Mercader (?) built a monumental Way of the Cross, with sculptures by Eusebi Arnau (b. 1854 - d. 1934) and Joan Pujol (?) that was financed by several different associations. These stations were destroyed at the beginning of the Spanish Civil War (1936). The only monument still standing is the Chapel of Our Lady of Solitude (1916), with paintings by Darius Vilàs (b. 1880 - d. 1950) and a sculpture by Josep Llimona (1864-1934), at the end of the way.

A new Way of the Cross was planned in the 1950s, with the architect Francesc Folguera (b. 1891 - d. 1960). There are two well-defined periods. The first, with sculptures by Margarida Sans Jordi (b. 1911) and Francesc Juventeny (b. 1906-?), and the second, with works by Domènec Fita (b. 1927). At the end of the 1950s, however, rock slides and little respect on the part of the public meant that the works of art had to be removed and replaced by a simple star designed by Josep Garganté (b. 1943). If today we can enjoy a striking view of Montserrat from the 12th station it is because early in the morning of Holy Thursday in 1991 a large rock fell and make the hole we can still see today.

To the right of the square in front of the Chapel we can see the steps that take us to St. Michael's Path. Once there, we can calmly return to the shrine, or continue on with one of the other suggested itineraries.

Brother Garí's Viewpoint

The way that takes us to Brother Garí's Viewpoint (810 m.) begins a few meters above the end of the Way of the Cross, before St. Michael's Path. It is marked, although the first steps are a bit hidden behind the trees. For many years this brief but pretty route was difficult to traverse, but today it is in fine condition. One only needs to be rather agile.

The 80-meter drop is easy to negotiate with the zigzagging path that goes up the Monkeys Channel, which often turns into a stream when it rains. The incline ends abruptly, and the trail becomes flat. We go in front of a cave where popular tradition has it that Brother Garí lived. Brother Garí is the mythical figure of the oldest legend about Montserrat, along with the legend about the finding of the Mother of God in the Holy Grotto. From Brother Garí's Viewpoint we can see a beautiful picture of Montserrat at the foot of a huge rock; in the distance we can see the rock formation known as the Elephant. We can also venerate the image of St. Mary, reproduced here in majolica.

St. Michael's Chapel

St. Michael's Path begins to the left of Abbot Oliba Square, and the path gently leads us to the chapel which lends its name to the route. In the first part of the journey we come across the little squares where pilgrims often meet or simply come to rest. Everywhere we see sculptures and statues commemorating some important event. The two most notable sculptures are those that represent Pablo Casals and St. Francis of Assisi. The former, a work of Joan Rebull (b. 1899 - d. 1981), is a monument (1976) to the great Catalan musician in commemoration of the 100th anniversary of his birth. A little further along, and after the camping area, the only part of Montserrat where camping is allowed, we see the slender statue of St. Francis, an offering made by the Franciscan Tertiaries of Catalonia in 1927 to commemorate the 700th anniversary of his death. It is the work of the sculptor Josep Viladomat (b. 1899 - d. 1989).

St. Michael's Gate (*La Porta de Sant Miquel*) that we cross later marks the end of the shrine area to the east.

The path we are taking was the most important way to Montserrat in medieval times; it departs from the town of Collbató, at the foot of the mountain. We, however, will only take the last part of the route, to St. Michael's Chapel, where after their long journey pilgrims saw the monastery and shrine for the first time, and organized themselves into a procession in order to show their devotion and respect for St. Mary.

About 100 meters before arriving at the Chapel we can take a path to our left which will lead us to St. Michael's Cross (770 m.), a magnificent viewpoint. You can see from the Pyrenees to the Llobregat delta, in addition to getting the whole view of Montserrat. An iron cross reminds us of the crosses of medieval and baroque prints crowning the hills on the mountains.

We go back to St. Michael's Path and then go to the Chapel (825 m.). It was formerly a hermitage, according to documents from the 10th century. On top of those foundations in 1870 the Abbot, Father Miquel Muntadas, had the Chapel we see today built. Due to the large numbers of visitors, the porched areas where added in 1958.

St. John via St. Michael's Path

If we are tired, or short on time, we can take the same path back to the shrine. If not, a little further along we come to St. Michael's Opening (*Pla de Sant Miquel*; 855 m.), where we come across three paths. We choose not to take either of the two to the left; one goes directly to Collbató, and the other, more narrow one would take us to the Holy Grotto. If we take the path to the right, known as the Path of the Hermitages, we will go to St. John's. Once there, if we pay attention to the schedules, we can take the funicular railway down, which leaves us where we began. For more information, please see Itinerary VI.

Itinerary IV

Apostles' Square (La Plaça dels Apòstols; 20 minutes)

Path of the Degotalls or Path of the Magnificat (45 minutes)

Apostles' Square

From Abbot Oliba Square we go down the road towards the exit. Once we have crossed the Square of the Cross (*Plaça de la Creu*), we go to the right and take the Boys' Choir Path (*Passeig de l'Escolania*) under a large awning, at the end of which we come across a building designed in 1975 by the architect Father Pere Busquets (b. 1925) with a restaurant. Before reaching that point, however, and if it is morning, we can see a small, typical market. Here sell their products the women from Marganell, a village on the mountain. The women come to Montserrat every day to sell their products; the most famous products here are honey and fresh cheese.

Once we have finished the Boys' Choir Path, we can go up the terraces of the building. From here, and looking back, we can see the typical picture of Montserrat: first, the rather plain 16th century basilica surrounded by buildings dating from various periods, corresponding to the Boys' Choir and the monastery; in the distance, we can see the lodgings for pilgrims, and even further in the distance we can see the best known peaks of the mountain: *Gorra Frígia* (1,152 m. in altitude) and *Magdalenes*, which tower over St. Mary's Stream, which goes down to the Llobregat river.

From these terraces, but now looking towards the mountain, we can see one of the most popular profiles: at the top we can see St. Michael's Chapel (*Capella de Sant Miquel*) and St. Michael's Cross at the edge that drops some 200 meters; below we see a winding path that takes us to another chapel, the Holy Grotto. For more information on these two chapels, please see Itineraries III and V.

Crossing the terraces now, and right next to us we see a rather curious architectural structure. Let's get closer. We see some openings separated by thin columns. If we turn towards the neo-Romanesque apses (1876-1885) attached to the church's façade (16th century). This was what used to be the third floor of the central apse of the church, which ended in an open gallery, lending volume to the church of that time. In 1992, during the restoration work, which basically consisted of returning the church to its original volume of the 16th century, the gallery was moved here. To better understand these changes, there is a plaque with a map to our right explaining this restoration (1991-1995). Under the small round square where we are standing there is a funeral chapel (1958) containing the remains of the members of the *Terç de Requetès de la Mare de Déu de Montserrat* who died during the Spanish Civil War (1936-1939). The project was designed by Father Pere Busquets and Father Crisòleg Picas (b. 1922). A little further below there is a monument to Raimundus Llull (b. 1232 - d. 1316), the mystic and Catalan writer from Majorca. It represents the Ladder of Knowledge, a reference to the work of this great medieval philosopher. The monument (1976) was designed by Josep M. Subirachs (b. 1927).

Once back in the little round square we can now say a few words about Apostles' Square, a remembrance of a chapel that was built here in the 16th century in memory of the Apostles. Partially destroyed in the Napoleonic Wars, it was rebuilt in 1858 and then restored in 1907. It was torn down at the beginning of the Spanish Civil War (1936), and the area was landscaped. On October 7, 1939, Abbot Marcet placed the first stone of the funeral chapel, the construction of which was not undertaken until 1958. Despite all of this, the name of the chapel has come down to us. Today much of the area is used for parking.

Path of the Degotalls or Path of the Magnificat

Once we have crossed Apostle's Square, we can go to the sidewalk to the left of the road, bordering on a large stone wall. Right in the middle there are two sculptures (1960) representing St. Benedict and St. Scholastica, the work of the sculptor Jaume Clavell i Nogueras (b. 1914 - ?). Going further towards the exit, we see to our right a large monolith with a cross, known as the Pax vobis (1960), designed by Father Pere Busquets (b. 1925), which marks where the shrine used to begin. To the left we see a sign *Els Degotalls-Camí del Magníficat*. Going up a ramp to our left, we see the path. We can first go up a short way to see the bronze sculpture the Bethlehem Society of Barcelona erected in memory of the Catalan poet Father Jacint Verdaguer (b. 1845 - d. 1902) in 1931. Among several poems dedicated to Montserrat, he wrote the lyrics to the Virolai which is sung after midday mass. The sculpture is the work of Carles Flotats (?-1950), and it has been located here since 1953, when Artists' Path began in the first part of the Path of the *Degotalls*.

This monument is followed by others in commemoration of illustrious people from our past. The second is a monument to the writer Joan Maragall (b. 1860 - d. 1911), placed by the Friends of Joan Maragall Society on July 5, 1958 with a project designed by Frederic Marés (b. 1893 - d. 1962). Attached to the rocks we see a plaque dedicated to the tenor Emili Vendrell (b. 1893 - d. 1962), placed to commemorate the 100th anniversary of his birth; it was designed by the jeweler Joaquim Capdevila (b. 1944). A little further on we see the monument to Pep Ventura (b. 1817 - d. 1875), the composer of hundreds of *sardanes*. Every year a commemorative key is placed on it. It was designed by Josep Mainar (b. 1899 - ?) and Alexandre Cirici i Pellicer (b. 1914 - d. 1983), and the sculptors Rafel Solanic (b. 1895 - d. 1990) and Carles Collet (b. 1902 - d. 1983) did much of the work. Amidst the trees, on a rock, there is a bronze medallion dedicated to Pompeu Fabra (b. 1868 - d. 1948), the person responsible for codifying modern Catalan, placed on the occasion of the 100th anniversary of his birth. Next to two fountains there is a monument to the musician Anselm Clavé (b. 1824 - d. 1874), designed by Ferran Serra i Sala (b. 1905 - ?) in 1955. Five minutes further along there is a beautiful, smooth plaque with a single frieze, designed by Joan Rebull (b. 1899 - d. 1981), that the Orfeó Català offered in memory of its founder, Lluís Millet.

The path continues on for about 15 minutes around the mountain; it is flat and shaded. We can see the cliffs close-up; they fall some 200 meters from the path where we are standing. You can also see the Plain of Bages and, on clear days, the Pyrenees from here.

After the part dedicated to artists, we now come across some 60 majolica statues, a representation of the worship of the Catalan people for Mary, Mother of God. Most are the work of Joan Guivernau (b. 1909). The oldest series of this collection is in the middle, and has a text known as the *Magníficat* or Canticle of Mary. At the end we come to a small square with no exit; to the left there is *Degotalls*, a rock blackened by the water that used to drop between the rocks.

The two forks to the right we will have passed would take us to the Can Maçana road (please see Itinerary VIII).

ITINERARY V

The Holy Grotto
(45 minutes)

Like practically all the routes we suggest, this one also begins in Abbot Oliba Square. There are two options. The first is to take the funicular railway to the Holy Grotto, inaugurated in 1929. The station is almost at the beginning of St. Michael's Path and lets us off in the square where the monumental Rosary begins. This route eliminates many of the stairs that might be an obstacle for older visitors, as the rest of the route is flat. The second possibility is to walk there, which is what we describe below. This route runs 1.5 kilometers, and in either case we need to get the information as to when the Holy Grotto is open and when the funicular railway returns. The Information Office will provide free-of-charge the brochure on the Holy Grotto, as it is a nice complement to this guide.

The Path to the Holy Grotto and the monumental Rosary

From Abbot Oliba Square let's walk towards the Square of the Cross. We cross it, and begin going down towards the exit. To the right of the sign there are some steps. Right before them, there is a sign: *Santa Cova.* After the first series of steps we go past the Cable-Car, which has been in service since 1930. Here to the

left we can see the old train that went through a tunnel. That line was in service 1892-1957. We continue on the wide steps to the right until we come to the square where the Monumental Rosary begins. Halfway down we pass the sculpture of St. Dominic (1970) by Josep M. Subirachs (b. 1927). To the right of the square we can see the lower station of the funicular railway, which, schedules permitting, we can take if we don't want to walk up all the stairs.

It is worthwhile to spend a minute and think about the difficulties, time, and economic costs for our ancestors in building this path. Work began in 1693 and had to be finished just one year later. But there were serious problems, arising from the need to deal with the severe drop in the mountain. It should come as no surprise that work was not finished until 1704.

This route, with its surprising views, has been made even more beautiful by the 15 groups of sculptures corresponding to the mysteries of the Rosary. They are to be found all along the path to the Holy Grotto.

With the financial help of associations, institutions, and families, this Monumental Rosary began in 1896 and was completed in 1910. Many architects, sculptors, and artisans took part; that is why each part has its own style. The decisive role of architects like Josep Puig i Cadafalch (b. 1867 - d. 1956) or Antoni Gaudí (b. 1852 - d. 1926) and of sculptors like Josep Llimona (b. 1864 - d. 1934) or the Vallmitjana brothers (b. 1828 - d. 1919, b. 1830 -d. 1905) has unified it. It is rightly known as the most important architectural monument in the outdoors of the Catalan modernist movement. In 1983 a new second mystery of joy was needed, and a new bronze sculpture by Manuel Cusachs (b. 1933) was put in place.

Once we have reached the fifth mystery of sorrow, the Crucifixion of Jesus, which we will recognize from the bronze modernist cross designed by Josep Puig i Cadafalch which stands in front of the cliffs, which at this point drop some 400 meters, the path changes direction and becomes more open. It quickly leads us to the Chapel of the Holy Grotto (605 m.).

The Monumental Rosary of the Path to the Holy Grotto prepares us spiritually for the visit. The rosary, a devotional practice dating from the 12th century, forces us to think about 15 episodes or mysteries in the life, passion, death, and resurrection of Jesus in which Mary, His mother, was present. The Our Father and ten Hail Marys are recited between mystery and mystery, and at the end a *Gloria Patri* is said. This rite of devotion to Mary can be substituted by readings of some sections of the Gospel according to St. Luke which narrate the life of Jesus.

The Holy Grotto

The small area of the Holy Grotto has a railing around it, and we can rest in the square while we read about this place, which has been worshipped for centuries by so many pilgrims.

The legend of the finding of the Image of the Mother of God

The legend — the oldest text that refers to it is from 1239 — tells us that in 880 on a Saturday near dusk some shepherd children saw a great light fall from the sky, together with a beautiful song, in the middle of the mountain. The following Saturday they went there with their parents, and the vision occurred again. The four following Saturdays they went with the rector from the town of Olesa, and everyone saw the same vision. Once he found out about it, the bishop of Manresa organized a trip, also on a Saturday. There they saw a cave in which they found the image of St. Mary. They tried to take it in a procession to the city of Manresa, but they were thwarted, and that was understood as divine intervention: that Image had to be worshipped at Montserrat.

Whether in popular legend or explained by a trained historian, the fact is that the place is holy and has been a site of worship for many centuries, due to the Image of St. Mary and the pilgrimage of thousands and thousands of men and women who have come to pray and ask Mary to intercede for all of us. Thus Mary's words in the Gospel come true at Montserrat: "My soul proclaims the greatness of the Lord and my spirit exults in God my saviour; because he has looked upon his lowly handmaid.

As we walk in silence to respect this holy place, we see a chapel in the form of a cross supported by a rock formation on the mountain, which is where the original Holy Grotto was. At this very place, to the right of the alter, there is a stylized reproduction of the Holy Image, today located in the basilica. We may stop for a moment of worship and silence.

Yes, from this day forward all generations will call me blessed, for the Almighty has done great things for me." (Luke 1:46-49). We can all sing with Mary and through Mary, that God has looked at us, He loves us and saves us. This is the light, the guide of our lives, that we find here.

The Chapel of the Holy Grotto

The first pleasant surprise we have as we near the Holy Grotto is its immense beauty, the harmony between nature and architecture. They blend together to form one. Without exaggerating, we, like the shepherds who saw the light, are captivated.

The chapel dates from the end of the 17th century and the beginning of the 18th (1696-1705). We should pay attention to the very high supporting walls, which were necessary due to the slope of the mountain.

Now that we can visit the chapel, we see that there is a hemispherical dome over which there is a light to illuminate the area. This main chapel is connected to another construction next to a small cloister, with the ex-vots, sacristy, room for pilgrims and home of the monk who meets pilgrims, and a garden. Some of these areas can be visited.

This area was greatly damaged during the Napoleonic Wars (1811-1812), just like the monastery and shrine. Francesc de Paula del Villar i Lozano (b. 1828 - d. 1901) planned (1857) the restoration project, as the main walls were entirely uncovered. This project is basically what you see today, with the addition of some new elements. Work was begun in October, 1858 and finished in December, 1859.

The modern history of the Holy Grotto has several important points, but this is not the place to explain all of them. We will only mention the most important architectural points. On July 4, 1994 a forest fire destroyed the roofs next to the chapel and cloister, and this resulted in damage to the flooring, furniture and other open areas, which were also damaged by fire. The heavy rains of the following autumn then caused massive rock slides on the path and access to the Holy Grotto, as there was no plant life to support the rocks. The path and the Holy Grotto were closed for reasons of safety. A project to restore the Grotto was designed in 1995 by the architect Arcadi Pla i Masmiquel (b. 1945) and work was begun in 1996. There was another natural disaster, however: heavy rainfall on September 5, 1995 moved the skylight in the dome, as

the base must have been damaged by the fires of the year before, and fell on top of the dome, and everything caved in. The same architect did a second design, and the completely restored Holy Grotto was again opened to pilgrims on March 19, 1997.

Itinerary VI

St. John and the mountain chapels
(1 hour)

There are reasons to believe that the first men at St. Mary's Chapel, before the founding of the monastery in 1025, were hermits. When the monks from the monastery at Ripoll came to Montserrat, this hermitic presence did not disappear, but rather became consolidated. The hermits were a part of the community of monks living in the monastery, and also obeyed the priors or abbots of Montserrat. One piece of evidence for this close spiritual and legal tie are the "Constitutions" and rules for living to obtain spiritual perfection for hermits that were written by Abbot Garcias de Cisneros (b. 1455 - d. 1510). These rules were observed from the end of the 15th century to the beginning of the 19th century, thus allowing a prosperous hermitic lifestyle at Montserrat. The hermits became renown for their exemplary life. However, this way of life greatly suffered the

consequences of the Peninsular War against Napoleon (1808-1814), as several hermits were killed and the hermitages, like the monastery, were destroyed or abandoned, some forever more.

The increased number of hermitic monks brought with it the organization of this way of life. Thirteen hermitages were built at the top of the mountain, strategically separated from one another, with their respective paths and beautiful scenery. Today, most of these chapels are in ruins, silent witnesses to a unique institution in history. In some cases, another chapel was built where once a hermitage was. The name has been kept, and it corresponds to a saint's name. All this makes for an attractive stroll or, in the more remote areas, an excursion.

Today, with just a few exceptions, the hermits and their chapels are no longer there, but the scenery and landscape, rocks and forests, are still the same and preserve these men's solitude turned into prayer. We too can enjoy this solitude.

The two itineraries below go to two hermitages chapels where newer chapels were built at the end of the 19th century: St. John's Chapel and St. Jerome's Chapel. Once at St. John's Chapel or in the path to St. Jerome's, we can point out where other chapels were.

To St. John's Chapel (Sant Joan) on the funicular railway

We begin our trip from the Abbot Oliba Square (alt. 710 m.), as before. To go to St. John's (976 m.), we have three choices. We will begin here with our recommendation: taking the funicular railway. The other two choices are not difficult, but do require more time.

The lower station of the funicular railway to St. John's is to the right of the beginning of the *Camí de Sant Miquel* (St. Michael's Path). This railway was built in 1917 in the middle of the *Torrent dels Avellaners* (Hazelnut tree stream), and the ride to the upper station takes 7 minutes. The cars were completely remodeled in 1997 to make the panoramic views available to more visitors.

Once at the top, we are in a small square known as *Pla de les Taràntules* ("Tarantula Square"). From here we can take several different paths, the most important of which takes us to St. John's Chapel (20 minutes).

Before beginning this path, however, we suggest you visit the exhibit in the upper station building on features of the geology, flora, fauna, and prehistory of the mountain of Montserrat. It also contains information on the history of the monastery and shrine.

In this building there is also a marvelous terrace that provides us with a breathtaking view of the monastery and shrine, right in the middle of the mountain, protected from the north wind but open to the warm sun of the east.

St. John's Chapel

We now take the path beginning in the *Pla de les Taràntules* ("Plain of the Tarantulas") and slowly climb up the *Torrent Fondo* ("Deep stream"). Some parts are inaccessible and thus constitute one of the vegetation reserves of the mountain. Unfortunately, a forest fire in 1994 destroyed much of the forest in this area; that is why there are no oak (*Quercus ilex*) trees and the path has few areas of shade. Nevertheless, the beauty of this area is captivating, with its depth, crevices, and shapes, and the colors of new vegetation.

As we stroll, after a few minutes we will see St. John's Chapel in the distance (alt. 1,025 m.), at the top of a hill next to the path. It is a relatively modern chapel (1893) whose structure is relatively uninteresting. However, from this point, to our right, we can see some very interesting buildings. With the exception of the building to the extreme left (a former restaurant), we can see two small buildings seemingly hanging from the rocks. They are the hermitages known as

St. John and St. Onofrius, situated in a lonely place that still stands out today. In past centuries they were highly valued for their exceptional location. St. John's hermitage, from which the area gets its name, was chosen by some abbots of Montserrat to spend their last days. We know that King Philip III visited it on July 10, 1599, amongst many others. Abbot Garcias de Cisneros (15th century) spent long periods at St. Onofrius hermitage, with its splendid view.

Under St. John's Chapel, and following a narrow path that curls downward, we arrive at St. Catherine's hermitage, a large part of which is a cleft in a large rock. Today the vegetation has reached the end, and only goes to accentuate even more the isolation that characterized this point. It is said that the most frequent visitors the hermit monk had were birds.

Even though we know these monks lived hermitic lives, we know they were visited by others. Often the visitors were pilgrims who, once at Montserrat and after their stay, wanted to have some contact with these men of God. The hermits listened to them and

gave them advice, but more than anything else provided a strong point of reference. With their inner peace and conviction, these men devoted to prayer, penitence and work were looked up to.

One of their jobs, in addition to taking care of crops and studying, was to sculpt briar. Briar was used to make spoons, glasses, bowls, rosaries, crosses, etc. because it is hard and compact. The briar crosses made by the hermits at Montserrat are famous and can be recognized by the traditional symbols related to the passion and death of Jesus they engraved on them, in addition to the presence of the coat-of-arms of Montserrat.

Visitors took them as a symbol of the serene way of life they had shared, and they were always a token of their experience and a challenge for their own lives.

Following the *Camí de Sant Joan* (St. John's Path) we will reach the end, and then taking a smaller path a few meters further we reach *Miranda de Sant Joan* (St. John's Viewpoint), from which we can see the southern side of the mountain, the Agulles area, which gets its name from the jagged edges of the hills. You can see beyond the Penedès to the area of Montsant.

We can go back to the *Pla de les Taràntules* ("Plain of the Tarantulas"; the upper station of the funicular railway) by the same path. Once there we have several choices. The first is to take the cable-car back to the monastery and shrine. The second, time permitting, is to go back via the *Camí de les Ermites* (Path of the Hermitages), which will take us to St. Michael's Path (see Itinerary III - to St. John via St. Michael). This is the path to the left of the *Pla de les Taràntules*, if our backs are facing the station. It is a wide, well kept path that in a short time allows us to see this part of the mountain, which has no peaks. The view always makes an impact on those who see it for the first time. When the path is directly over the station, at 999 m. in altitude, we have an excellent view of the western part of the mountain (1,236 m.), *Montgròs*, and the nearer peaks of *Gorra Marinera* (Seaman's Cap) and *Sant Salvador*. This route is a little over 3 kilometers to arrive at the monastery. The third possibility is to take the path that takes us to St. Jerome part-way or all the way up to the highest peak at Montserrat (see Itinerary VII).

To St. Joan via the Stairway of the Poor

At the beginning of this itinerary, we said there were two other ways to reach St. Joan in addition to the funicular railway. We can either take the Stairway of the Poor (*Escala dels Pobres*) or St. Michael's Path (*Camí de Sant Miquel*).

We will only briefly describe here the Stairway of the Poor route, as this route would be an excursion for many travelers and that is not the main purpose of this guide.

We reach the Stairway of the Poor, which gets its name from the fact that centuries ago there was a house for the poor and vagabonds, once we have climbed the stairs to the left of the fountain in Abbot Oliba Square. At the end there are signs for two paths. The path to the left leads us to the *Via Crucis* (see Itinerary III). The path to the right is the one we want to take. It goes along

St. Mary's Stream (*Torrent de Santa Maria*) and climbs upwards. Once we have finished going up the steps, there is a crossroads of paths. Now take the path to the left, which starts out flat but soon rises once we have gone past the remains of St. Ann's hermitage and have crossed St. Mary's Stream. There are three other crossings; to go to St. John always take the path to the left.

Another option is to take the path straight ahead of us once we get to the crossing at the end of the Stairway of the Poor. We will go along the side of a large rock, known as the *Panxa del Bisbe* ("the Bishop's tummy"), before reaching the *Pla dels Ocells* ("Birds' Point"; 930 m. in altitude). Take the path to the left, which will then take us past a solitary rock known as *Trencabarrals* ("Jugbreaker") to St. John's path.

To St. John via St. Michael's path

The last possible choice is to go to St. John via St. Michael's Path, which we described in Itinerary III of this guide. This is a pretty stroll that is not difficult, but does require time (1 hour, 15 minutes). If you wish to take this trail, do as we suggested above: first take the funicular railway to St. John and then walk the Path of the Hermitages to reach St. Michael's Path, which will eventually lead us to Abbot Oliba Square (50 minutes).

Itinerary VII

The Path from St. John to St. Jerome
(1 hour)

Even though we are about to describe here the walk up to the highest point on the mountain of Montserrat (1,236 m. in altitude), no heroic task is involved. Rather, from St. John's, and since most will have taken the funicular railway up, the path that takes us to our destination is relatively flat. With just a few exceptions, it is only in the last 10 minutes that the path is really uphill.

Our walk begins directly to the right of the *Pla de les Taràntules*, at the upper station of the funicular railway. Just a few steps away from here, the path winds between rocks and cliffs, and there is little fear of vertigo because the forest acts as a natural railing. We can still see the monastery and shrine; everything looks little, some 300 meters below.

Significant rocks

We take the path, staying to the left, and at the first bend we can see the basin of St. Mary's Stream (*Torrent de Santa Maria*), towering over which there are immense rocks. Each one has its own popular name depending on what its shape suggested. So, on the other side, in front of us going from right to left, we can let our imaginations go and make out rocks with the following names: The Little Mummy, The Mummy, The Cat, The Elephant, and the highest peak, St. Salvador (1,152), which is the name of the hermitage on this rock. There are documents referring to a hermitage here dating from as early as the 13th century.

Hermitages

If we look down about 100 meters, we can see St. Benedictine's Chapel between the live oaks (*Quercus ilex*). It was built in 1927 on top of the remains of the old hermitage of the same name. This was the 'newest' hermitage, as it was built by Abbot Pere de Burgos in 1536. This view, with the rocks, forests, and the chapel in the middle, is one of the most idyllic and suggestive of the peace that has characterized this place, devoted to the hermitic life, for hundreds of years.

Before going on any further, let's look up. To the extreme right of the cliffs in front of us, we can see the hermitage of *Sant Dimes*, which was inhabited as early as the 15th century. Close to it is the hermitage of the Holy Cross; located directly above the

monastery, it is practically removed from everything. Six-hundred steps lead up to it, and most of them open out to the rocks. In the past thirty years there has been a monk from the monastery living there as a hermit.

Right above the hermitage of the Holy Cross we can see the Plain of the Trinity (*Pla de la Trinitat*), where there are the remains of the hermitage of the Holy Trinity. At the end of the 15th century it was restored by Father Bernat Boïl, who later would accompany Christopher Columbus on his second voyage to America (1493) with the title of the first Apostolic Vicar of the West Indies, given by Pope Alexander VI. In 1625 Abbot Beda Pi made it even larger and made it into the largest of the hermitages. It was occupied until just before the destruction of Montserrat by the French (1811-1812). Father Gaspar Soler was killed there on April 24, 1822 and was thrown into the water tank by outlaws. That was when the monks at the monastery decided to not allow the other four hermits to continue living there on their own. That was when the hermitic way of life, which had been so enriching for hundreds of years, came to an end at Montserrat.

We can now continue on next to the live oaks (*Quercus ilex*), which shade us until we reach the four points that are so characteristic of Montserrat that they appear in all the pictures of the mountain taken from St. Mary's Square. They are named, from right to left: *Gorra Frígia* (1,152 m.), *Magdalena Superior, Magdalena Inferior, Gorra Marinera*. There are the ruins of the St. Magdalene hermitage (15th century) on *Magdalena Inferior*, but perhaps the most interesting sight is that of St. James' hermitage seemingly hanging in the middle of *Gorra Marinera*. The hermits surely had a beautiful view of the monastery, but they must have been cold in the winter, with the cold air whipping through the mountain.

The mountain's vegetation

The path continues along St. Mary's Stream. At various points we can see *Cavall Bernat*, and in the distance, St. Jerome. When the path crosses the torrent, we take some steps that will take us away from it for a while, but we will cross it again later on. We can stop at the stairs to rest a bit, take a deep breath and admire the scenery. We don't have to look far to see many of nature's wonders: the magnificent rocks behind us, in the middle of which there is a pine tree! And we thought that kind of image only existed in the imagination of painters.

Back at the torrent, we now see that we are actually walking in the middle of it. Fortunately for us, these streams only carry water when it rains hard, when they can be quite dangerous. We can see the old path next to the river bed we are walking in. The uphill walk is nice, under the live oak trees, and soon we reach St. Jerome's Chapel. If we pay attention to these live oaks (*Quercus ilex*) that we have seen from where we started at St. John's, we will have noticed two important things. One, they live at 700 m. - 1000 m. in altitude, which is what botanists identify as live oaks with briar (*Quercetum ilicis*

galloprovinciale, subsp. viburnetosum lantanae), along with a dozen other species, shrubs and herbs. This type of oak is different from the oaks we find close to the shrine (*Quercetum ilicis galloprovinciale*). The other thing we see is that these trees are not hundreds of years old, as might be expected on this part of the mountain. We have to bear in mind that until the beginning of the 20th century the trees on the mountain of

Montserrat, like most of the forests here, were periodically cut for firewood, to heat homes, to use in ovens, and to run small industries in our cities and towns.

St. Jerome's Chapel

St. Jerome's Chapel (1,130 m.), which we reach at the end of an uphill climb eroded by the rain, was built in 1891. The area to the right has been covered to protect visitors from sudden storms. It is better to stay here, as opposed to trying to go down the usually dry river bed.

St. Jerome's hermitage was not located where the chapel is today, but rather was a little higher up. Two paths begin in front of the chapel. The one to the right will take us to the telecommunications tower the Center for Telecommunications *of the Generalitat de Catalunya* built in 1996 to eliminate the thirty-some antennae that were here while maintaining modern telecommunications for the area. The path to the left takes us to the highest part of Montserrat, the peak of St. Jerome (1,236 m.).

Walking up to St. Jerome

Let's take the path to the left, which goes directly up to the top. We saw this practically from the beginning. We first pass by an area where we can see that there were some buildings, but now only the water tank of what used to be a restaurant remains. Here is where the St. Jerome hermitage was. It was rebuilt in 1590, only to be destroyed again in 1811. To the left, on top of a rock, there is a point to see the scenery; it was built in 1956 and has a plaque that hikers had placed in honor of Father Jacint Verdaguer (b. 1845 -d. 1902) a few years earlier, to commemorate the 50th anniversary of this Catalan poet's death.

As we go up we can see the changes in vegetation, even in comparison with that around the chapel. We are now in a new zone of vegetation, with mountain live oak (*Quercetum mediterraneo-montanum*) that live in high areas of the mountain, between 700 and 1200 m., that have little slope. Here the rain can penetrate into the soil, and this results in a noticeable drop in the lime content of the substrate. Thus, some plants that flourish in acidic soils can live here, as opposed to in the other oak ranges. In addition, the shallowness of the soil and constant exposure to wind makes for few trees and

shrubs. Conversely, small plants such as goldenrod (*Solidago virgaurea*) and other indigenous plants such as *Veronica officinalis, Prunella hastifolia, Fragaria vesca and Stachys officinalis* are common.

We are now at the highest point on the mountain of Montserrat, St. Jerome (1,236 m.).

The first thing we notice is the cool breeze. There is not a lot of room up here, but the area is protected by a metal railing. On one side we can see the large stainless steel compass that was placed by the Hiking Club of Bages (*Centre Excursionista del Bages*) in 1986, so that we can determine which direction is which. It also gives us the names of the areas we see on the horizon, and the most important sights we can see on clear days.

The view from atop St. Jerome

Now we can see that the mountain of Montserrat is privileged, even from a geographical standpoint. Our mountain rises 1000 meters from the plain, which means that from this point, which is not that high up, we have a splendid view of 360° with no immediate obstacles. We can see the Pyrenees to the north, to the west, the province of Lleida and the Aragonese

Pyrenees, to the south, Tarragona and the Mediterranean. It is even said that on very clear days you can see Majorca. And, to the east lie Barcelona and Girona.

Now, if we turn our backs to where we just came from, to our right we can see the basin of St. Mary's Stream, surrounded by mountain peaks and rocks. The shapes of the rocks and groves of live oak (*Quercus ilex*) are really quite captivating. Closer to us we see the telecommunications tower (1996), which is 45 meters high and stands where the old cable-car station was (1929). This station was dismantled in 1986. Standing by itself behind that is the rock known as *Cavall Bernat* (1,103 m.), on top of which there is an image of the Mother of God of Montserrat.

As we get closer to the railing, we can see the tremendous cliffs that fall some 600 meters. At the bottom of these cliffs, right next to the *Can Maçana* road, there is an area surrounded by a garden. This is St. Cecilia, a Romanesque church dating from the 11th century (please see Itinerary VIII). A little further beyond to the right, near the road that goes from Monistrol to Montserrat, we can see a bell tower surrounded by some buildings. This is the Monastery of St. Benedictine, a new monastery for Benedictine nuns that was built in 1954.

To our left, the rocks continue to suddenly stop at the forests of the town of Bruc to the north. To the southwest we see the olive trees the surround Bruc; this is where the wild goats live. The rock formation right in front of us is known as The Echoes (*Els Ecos*).

Before going back, we can stop to look at the cliffs again. In the nooks and crannies we can see some very pretty wildflowers (*Saxifraga callosa and Ramonda myconii*) that manage to live in extreme weather conditions, where most plants would not survive at all.

At the top of this peak there was once a chapel in honor of St. Mary, known as St. Mary the

highest (*Santa Maria la més alta*), as if to say that from the highest hermitage, St. Mary protected the mountain and all the land commended to her. This is a good moment for a short prayer of thanks to God for nature and all the wonderful and beautiful things we see and have during our lives.

Itinerary VIII

The walk to St. Cecilia

It is about four kilometers from Abbot Oliba Square, where all of our trips have begun, to the Church of St. Cecilia. We can either go by car, or it can be easily reached on foot. Either way, we must begin our journey at the exit on the road to Montserrat.

If we are walking, we take the *Camí dels Degotalls* (Itinerary IV). After walking for about 10 minutes, we will see the majolica image of the Mother of God of Gleva (Vic), a little further up there is a single cypress tree right at the beginning of a path to our right that leads directly to the main road. Take this road, and a little further along, in a flat area close to the road, we see on the other side a little square with trees and stone benches. This little area is known as *St. Jaume el Blanc* (St. James the White); the name dates from at least as far back as 1532. It is related to the passing of pilgrims on their way to Santiago de Compostela in Galicia, after they had stayed at St. Mary of Montserrat.

Once we are on the road to Can Maçana, which would take us to the road connecting Barcelona to Lleida, we can take a quiet stroll next to the road while looking at the highest mountain peaks rising to our left.

Quite abruptly, in a bend in the road, we see the striking rock formation known as *Cavall Bernat* (1,103 m high) jutting out. Just a little further along we see that we have left behind the group of hills known as *Flautats*, in front of us lies the *Paret dels Diables* (Wall of the Devils), which is overlooked by the Chapel of St. Anthony, which we cannot see.

A little before we reach St. Cecilia's we can also see the telecommunications tower of St. Jerome (1,200 m), which rises above the huge *Paret de Sant Jeroni* (St. Jerome's Wall). The enormous proportions are truly impressive, not only because of the change in altitude of over 600 meters but

also because these cliffs are part of a mountain and an area that is relatively small.

There is a short tunnel in the road, and then there is a slight incline up to St. Cecilia (675 m). The first thing we see to the right is a group of tightly packed cypress trees, in the middle of which there is a mountain refuge belonging to the Catalan Federation of Excursion Groups, where mainly mountain climbers stay. Given its characteristics, the mountain of Montserrat has become a major draw for mountain-climbers in the past few decades.

The Monastery and Church of St. Cecilia

A few meters away we see the Romanesque apse of the Church of St. Cecilia and the surrounding buildings. The first reference we have of St. Cecilia goes back to a document from 945, which tells us that a monastery had been founded. St. Cecilia is therefore the first place on the mountain where a religious community was established, although the community here never grew to be very numerous. It became a part of the Monastery at Montserrat in 1539.

The Church of St. Cecilia, with three naves of varying lengths and three arched apses, is a magnificent example of Romanesque architecture from the 11th century. In 1811 it was raided and partially destroyed by invading French troops. It was partially restored and reopened as a church on November 22, 1862.

It was not fully restored until 1931, under the direction of the famous architect Josep Puig i Cadafalch (b. 1867 - d. 1956). The church is generally open during daytime and can be visited. The buildings next to the church date from various periods and have been used for different purposes over the years. As of the 1960s the Monastery has used them as lodging for young people.

St. Cecilia's arboretum

The four hectares of land surrounding St. Cecilia have not been cultivated for many years, as can be seen in the first photographs of the monastery. After the forest fire of August 18, 1986 destroyed the vegetation on the mountain's north face and the surrounding forests, this area has been replanted as an arboretum in an attempt to

reforest the mountain. Today there are some 650 trees and 1,300 shrubs belonging to 150 different species, as a remembrance of what must never happen again and that we all are responsible for the ecological heritage that has been entrusted to us.

The arboretum is divided into two areas by a wide path. The area to our right is devoted mainly to reproducing the various types of oaks (*Quercus ilex*) found in Mediterranean climates, with many deciduous trees and other shrubs. There are two scenic points from which to view the Pyrenees and the Bages Plain, with the city of Manresa in the center. As we look out, the first thing we see is the Monastery of St. Benedict (1954), a convent for Benedictine nuns designed by the architect Lluís Bonet i Garí (b. 1893 - d. 1993). To our

left, the arboretum contains many evergreen trees and shrubs, most of which are native to the mountain. If we look at them closely, we see that one side grows better than the other, because the north wind is frequent and strong in this area and its effect on the vegetation is plain. From this point we can have a panoramic view of the rocky areas and peaks we have

mentioned in this itinerary. To the left of the mountain we see *Flautats*, then, *Cavall Bernat*, and then St. Jerome (1,236 m.) under which is St. Cecilia. To the west we see the *Ecos* (Echoes) area and the *Frares Encantats* (Enchanted Monks), and at the end of the mountain, the *Agulles* (Needles), which is beyond our view.

Just a few minutes from St. Cecilia, right alongside the road, there is a small, old restaurant that is one of the characteristics of the area. It serves traditional Catalan fare, and is open only during the summer months.

Hours

RELIGIOUS SERVICES

Weekdays and Saturdays

7:30 AM Morning prayer (Lauds)

9:00 AM Mass (May through October)

11:00 AM Conventual Mass

12:00 AM Mass

1:00 PM Salve Regina and Virolai sung by the Boys' Choir (except Saturdays from the end of June through mid-August)

6:15 PM Rosary

6:45 PM Evening prayer (Vespers) and Salve Regina (Monday through Thursday, sung by the Boys' Choir)

7:30 PM Mass (Saturdays and evenings before holidays)

Sundays and holidays

7:30 AM Morning prayer (Lauds)

8:00 AM Mass (May through October)

9:30 AM Mass

11:00 AM Conventual Mass

12 Noon Salve Regina and Virolai sung by the Boys' Choir (except Saturdays from the end of June through mid-August)

1:00 PM Mass

5:30 PM Mass (May through October)

6:15 PM Rosary

6:45 PM Evening prayer (Vespers), Salve Regina and hymns sung by the Boys' Choir (the Boys' Choir is on summer holiday from the end of June to mid-August)

7:30 PM Mass

Hours for veneration of the Virgin:

From **8:00** to **10:30 AM**

From **12:00 Noon** to **6:30 PM**

Also from **7:30** to **8:30 PM** on Saturdays and Sundays

THE HOLY GROTTO

Hours

Summer: (May through October)

From **10:20 AM** to **5:25 PM**

Winter: (November through April)

From **11:20 AM** to **4:25 PM**

Mass: Sundays and holidays at **11:30 AM**

INFORMATION

For more information, please go to the information office

Summer hours:
9:00 AM to **7:30 PM**

Winter hours:
9:00 AM to **5:45 PM**

FOR PASTORAL INFORMATION

Daily

9:30 AM to **1:00 PM**

4:00 PM to **6:00 PM**

OTHER SERVICES

Hotel: Open all year round

Museum: Open daily, **10:00 AM** to **6:00 PM**

Shops-Souvenirs: Open **9:00 AM** to **7:00 PM**

Cafe-Bar: Open **9:00 AM** to **7:30 PM**

Self-service cafeteria: Open **Noon** to **5:00 PM**

Restaurant: Open **Noon** to **5:00 PM**

St John's Funicular (*Funicular de Sant Joan*): **10:00 AM** to **6:00 PM** (departs every **20 minutes**)

Holy Grotto Funicular (*Funicular de la Santa Cova*): **10:00 AM** to **1:00 PM** and **2:00 PM** to **5:45 PM** (departs every **20 minutes**)

Cable car: 9:25 AM to **1:45 PM** and **2:20** to **6:45 PM** (departs every **15 minutes**)

Montserrat Rack Railway: 8:06 AM to **7:42 PM**

Please note that winter schedules differ substantially from summer schedules. During the winter months, we ask you to first contact our Information Office or check the web pages of Montserrat, the Montserrat Cable Car and the Montserrat Rack Railway for up-to-date information.

Services and installations

1 The Basilica
2 Our Lady
3 Centre for Pastoral Attention
4 The Museum of Montserrat
5 "Inside Montserrat" audiovisual presentation

Food

6 Restaurant Abat Cisneros
7 "Apostles" scenic viewpoint
- Cafe-bar del Mirador
- Self-Service
- Restaurant de Montserrat

8 Cafeteria

Lodging

9 Hotel Abat Cisneros ***
10 Abat Marcet Cells (*apartments*)

Shops

4 Museum Shop
5 Audiovisual Shop
7 "Apostles" scenic viewpoint Shop
11 Shop

To Sant Jeroni

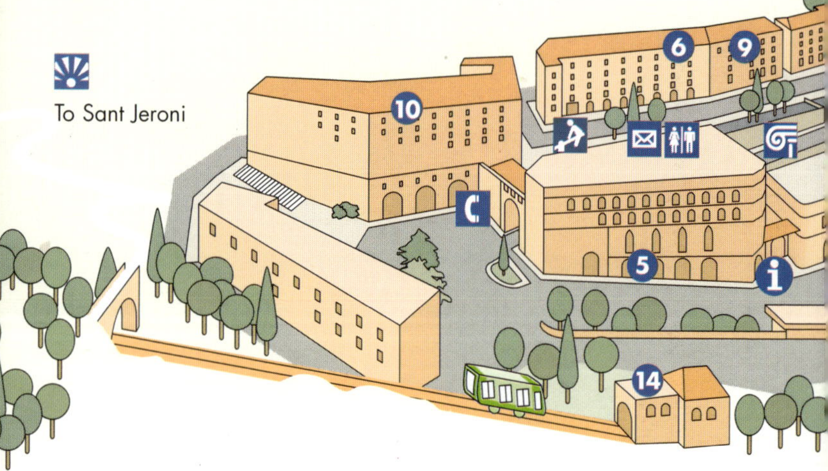

The Way of the Cross

Audiovisual Area

Restaurant Abat Cisneros

Museum of Montser